Aurora Simionescu

Magnetic Fields in Galaxy Clusters

GRIN Verlag

Imprint:

Copyright © 2005 GRIN Verlag GmbH
Druck und Bindung: Books on Demand GmbH, Norderstedt Germany
ISBN: 978-3-640-24413-3

This book at GRIN:

http://www.grin.com/en/e-book/90115/magnetic-fields-in-galaxy-clusters

Magnetic Fields in Galaxy Clusters

May 16, 2005

Aurora Simionescu

School of Engineering and Science
International University Bremen
Campus Ring 1
28759 Bremen
Germany

Contents

Abstract

An adaptive mesh refinement simulation of galaxy cluster formation was performed that included the passive evolution of a magnetic field. It was found that structure formation plays an important role in amplifying large-scale magnetic fields and that the magnetic properties of the obtained cluster were in good agreement with recent observations.

The initial field was amplified by a factor of up to 1000 during the formation of the cluster, and the field strength was seen to be well correlated with the gas density. We further found a magnetic energy power spectrum that is well described by -5/3 Kolmogorov-type turbulence. Near the accretion shocks on the outskirts of the cluster, the magnetic field is amplified well beyond the value expected from mere compression of gas. Here, shear flows lead to a substantial increase in field strength.

Realistic Faraday rotation measures were obtained from the simulation data, which was however not resolved well-enough to allow for a more quantitative analysis.

1 Introduction

Recently, magnetic fields in galaxy clusters have come to the attention of the scientific world as the largest-scale magnetic structures measured to the present date. While little is known about the origin of these fields and their evolution throughout cosmic time, the presence of large-scale magnetic fields may have important implications for the processes observed in galaxy clusters. Examples of these implications include inhibition of transport processes - such as heat conduction, spatial mixing of the gas, and the propagation of cosmic rays - or even dynamical importance through the Lorentz force and the additional magnetic pressure term. It is therefore of great scientific importance to determine the origin, evolution and structure of cluster magnetic fields and their relevance in structure formation and in astrophysical phenomena observed today in galaxy clusters.

1.1 Galaxy Clusters: Formation and General Properties

Clusters of galaxies are the largest gravitationally bound systems in the Universe. They can be recognized in the optical range as groups of galaxies which are located closer together than the average distribution. However, galaxies represent only around 5% of the total mass of a typical cluster. Most of the baryonic mass, roughly 20% of the total mass of the cluster, is contained in the form of hot ionized gas in the intracluster medium (ICM). The ICM is characterized by high temperatures (in the order of 10^8 K) and electron number densities in the range of 10^{-3} cm^{-3}; thermal bremsstrahlung emission from the hot gas in the intracluster medium is very intense in the X-ray spectrum, typical luminosities ranging between $10^{43} - 10^{46}$ erg/s. By far the largest contribution to the mass of galaxy clusters is the dark matter, which constitutes between 70-80% of the total mass and plays an important role in the development of the clusters. According to the model of hierarchical structure formation, clusters of galaxies are thought to form upon the gravitational merger of smaller units, such as groups and subclusters.

1.2 Theories Regarding the Origin of Cluster Magnetic Fields

Our Universe is permeated by magnetic fields on different length scales and strengths. The most palpable example is our own planet, which has a magnetic field of about 0.5 G at its surface. Moreover, the magnetic activity of our Sun also affects life on Earth. Magnetic fields in the Sun are about 10 G at the poles and can reach up to 2000 G in sunspots. Magnetic fields have been measured in the intraplanetary medium ($\simeq 50\mu G$), in protostars and neutron stars, in the intragalactic medium ($\simeq 5\mu G$ in the Milky Way). It comes as no surprise that also the intracluster medium is magnetized. Although the uncertainty in measurements is still significant, the observations are concordant to the existence of fields in the microgauss range having correlation lengths of the order of 10 kpc.

There are several hypotheses regarding the origins of cluster magnetic fields, namely the *primordial* scenario, the *protogalactic* scenario and the *galactic* scenario. According to the primordial scenario, the fields originate from the early universe, prior to recombination. Recombination is an epoch in the evolution of the universe (at redshift 1000) when the electrons and protons combined to form atoms, whereby the photons decoupled from matter and the Universe became neutral. The matter in the Universe was reionized at a later stage, possibly after the formation of the first stars. The primordial scenario involves several possible mechanisms for the generation of the magnetic fields prior to recombination. One such mechanism is the Biermann battery effect, which occurs when the gradients of the pressure and the electron number density are not parallel to each other, implying that the system is not in a static equilibrium. To restore the equilibrium, a thermoelectric current is generated, from which the magnetic fields originate. Other mechanisms imply local charge separation occuring during the quark-hadron (QCD) or electroweak (EW) transitions. Estimated values of these seed fields are in the order of $10^{-21}G$. If the primordial scenario is indeed true, the seed fields should be observable by means of anisotropies in the cosmic microwave background (CMB) and by their effect on nucleosynthesis. Current measurements of anisotropies of the CMB place upper limits of about 5 nG on the strength of potential seed fields.

The protogalactic scenario predicts the possibility that the most important growth in large-scale magnetic fields occurred during the reionization era, which corresponds to the early-stage of galaxy formation.

Lastly, the galactic scenario involves the interaction of the magnetized intragalactic medium with the ICM in order to generate cluster magnetic fields. Galactic winds and radio jets emitted by active galactic nuclei (AGN) are possible ways in which fields from the galactic medium can be transferred to the ICM. Moreover, the significant concentration of metals in the ICM suggests that a large fraction of it is of galactic origin. However, the fields in the ICM have large correlation lengths, comparable to the typical size of a galaxy, therefore a mechanism is required to arrange the galactic-originating fields on ICM correlation scales. One possibility in this sense is to have helical fields, which under the constraint of conservation of helicity tend to order on larger scales for a minimal energy configuration.

1.3 Measuring Cluster Magnetic Fields: Methods and Results

Apart from the questions posed towards the origin of the cluster magnetic fields, another important scientific research area regards their currently observable properties. There are three different methods which can be employed in order to determine the strengths of magnetic fields in the ICM.

1.3.1 Synchrotron radiation

The first of these methods makes use of synchrotron radiation luminosities. Synchrotron radiation is produced by relativistic electrons gyrating in a magnetic field due to the Lorentz force. The synchrotron emission spectrum shows a peak at a critical frequency which is proportional to the average magnetic field and the square of the Lorentz factor, γ, while the power emitted in synchrotron radiation depends linearly on the square of the average magnetic field and on γ^2. The energy of the relativistic electrons can be shown to be proportional to the synchrotron luminosity and to the $-3/2^{th}$ power of the magnetic field. The strength of the magnetic field is usually estimated by minimizing the total energy, namely the sum of the energy in relativistic particles and the magnetic energy. This condition is fulfilled when the two different energy contributions are roughly equal, which enables the calculation of the average magnetic field. Furthermore, the degree of polarization of the synchrotron radiation gives a measure of the field uniformity (linear polarization corresponding to a uniform magnetic field). From the synchrotron emission method, typical magnetic field strengths in the ICM are computed as $0.4 - 1\ \mu G$ [3].

1.3.2 Inverse Compton effect

Another method employed to measure the average magnetic field strength in clusters uses Inverse Compton (IC) radiation luminosities in addition to synchrotron emission. The inverse Compton effect consists in the scattering of microwave background photons by the relativistic electron population. As a result, the microwave background photons gain momentum from the electrons and are turned into X-ray or gamma photons. Given that the IC and synchrotron radiations both originate from the same relativistic electron population, one arrives at a proportionality relation: $\frac{L_{sync}}{L_{IC}} \propto \frac{u_B}{u_{ph}}$, where both the IC and synchrotron radiations can be measured and u_{ph} is the density of the CMB photon field which can be calculated, therefore one can solve for u_B, the energy density of the magnetic field. Results of the IC method give average magnetic fields of $0.2 - 1\ \mu G$ [3].

1.3.3 Faraday rotation maps

The Faraday rotation effect completes the set of methods currently used to measure cluster magnetic fields. The principle of this effect consists in the fact that a magnetic field induces different refractive indices for left-handed and right-handed circularly polarized beams. Linearly polarized beams can be decomposed into opposite-handed circularly polarized components. Upon passing through a magnetized medium, due to the different refractive indices, a phase difference occurs between the two components, which is equivalent to the rotation of the plane of polarization of the beam by an angle

$$\Psi_{obs} - \Psi_{int} = \Delta\Psi = \frac{e^3\lambda^2}{2\pi m_e^2 c^4} \int_0^L n_e\left(l\right) B_\parallel\left(l\right) dl \tag{1}$$

This is usually written as:

$$\Delta\Psi = \lambda^2 RM \tag{2}$$

where, comparing the two expressions, the rotation measure RM is defined as:

$$RM = \frac{e^3}{2\pi m_e^2 c^4} \int_0^L n_e\left(l\right) B_\parallel\left(l\right) dl \tag{3}$$

In practical units, this is written as

$$RM[\mathrm{rad/m^2}] = 812 \int_0^L n_e[\mathrm{cm^{-3}}] B_\parallel[\mu\mathrm{G}]\mathrm{dl[kpc]} \tag{4}$$

The "internal" angle of the polarization plane as the beam is emitted, Ψ_{int}, is assumed to be constant over all wavelengths. The angle of polarization at the observer, Ψ_{obs}, is measured as a function of the wavelength and corrected for the rotation effect due to the magnetic field in our own galaxy. Fitting a straight line of the form $\Psi_{obs} = f\left(\lambda^2\right)$ to the measurement data yields, according to (2), the rotation measure which contains information about the magnetic field.

In the context of measuring cluster magnetic fields, the rotation measure is determined for each point on a grid of celestial coordinates, thus creating a rotation map of the emitting polarized source as it is seen through the magnetized cluster medium. These two-dimensional rotation maps are analyzed, revealing information about the strength and structure of the magnetic field. Sources of extended polarized emissions are, for example, radio halos, relics and mini-halos [4].

Rotation maps project the three-dimensional magnetic field profile, combined with the electron number density profile, onto a two-dimensional plane by integrating along the line-of-sight. The most common method of disentangling the projection-effect from the RM and retrieving information about the average magnetic field in the cluster is to assume that the field varies on a constant scale, Λ_c. The cluster is modeled as a set of cells of side-length Λ_c inside which the electron number density and magnetic field strength are constant; the orientation of the field according to this model varies randomly from cell to cell. The scale on which the fields vary is usually closely associated with the correlation length calculated from the RM map. Results from RM measurements fall typically in the range of 1-10 μG assuming cell sizes of around 10 kpc.

1.3.4 Reconciling magnetic fields derived from the three methods

Compared to synchrotron and IC measurements, Faraday rotation map analyses give magnetic fields which are roughly one order of magnitude larger. Several arguments can be invoked in order to explain this discrepancy. Firstly, the cluster magnetic field may show a range of coherence scales, and the presence of highly correlated small-scale fluctuations can enhance the rotation measures and thus produce higher estimates of the average field strength. Secondly, an anisotropic pitch-angle distribution would weaken the synchrotron radiation relative to the IC emission, leading to an underestimation of the IC-derived fields. Also, if a large relativistic population is located in the weak-field regions, a large part of the IC emission will come from low magnetic field-strength parts of the cluster.

1.3.5 The RM debate

The scientific debate regarding the calculation of magnetic fields from rotation maps has become increasingly intense as new methods of analyzing RM maps are being developed. As discussed by Rudnick [2], the evidence up to date is insufficient to prove the fact that the rotation measures are due to the ICM and not to a thin thermal skin mixed with the relativistic plasma of the emitting radio source. Moreover, Rudnick cites polarization percentages of 0.9 ± 0.7 and 0.9 ± 1.0 in beams which were used to produce rotation maps, an uncertainty which is too large for the results to still be considered relevant. Analyzing the rotation map of the Coma cluster by removing sources embedded in the cluster and keeping only the rotation measures of background sources, Rudnick found that this cluster showed, within error limits, a zero averaged absolute value of the rotation measure, even close to the cluster center.

Counterarguments in favor of cluster magnetic field-generated rotation maps are presented in a study by Clarke et. al. [1] who performed a statistical survey of rotation measures as a function of the impact parameter of the emitting source with respect to the cluster center. A clear broadening of the RM distribuion toward small impact parameters (below roughly 1 Mpc) was found, justifying the conclusion that rotation maps are due to the ICM rather than being intrinsic to the emitting source.

1.3.6 Improved methods to determine the magnetic fields from rotation maps

Recently, Enßlin and Vogt ([5], [6]) developed a method of analysing Faraday rotation maps which enabled them to use the RM of Hydra A to calculate the magnetic field of $7\pm2\mu G$ at the cluster center, within the assumptions made about the most likely geometry of the field.

In the first stage, Vogt and Enßlin [6] assume an isotropically distributed

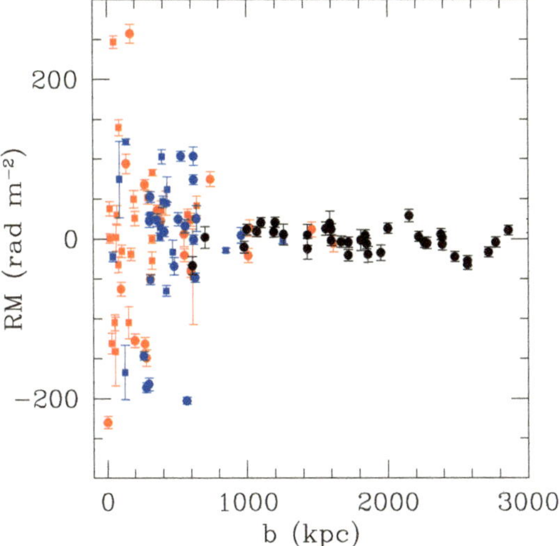

Figure 1: RM distribution as a function of the impact parameter from Clarke. Embedded samples are represented in red, background samples in blue and control samples in black

magnetic field for which they define a scalar autocorrelation function:

$$w\left(\vec{r}\right) = < \vec{B}\left(\vec{x}\right) \cdot \vec{B}\left(\vec{x}+\vec{r}\right) >_{\vec{x}}$$ (5)

Similarly, a scalar autocorrelation function can be defined for the rotation map as:

$$C_{RM}\left(\vec{r}\right) = < RM\left(\vec{x}\right) RM\left(\vec{x}+\vec{r}\right) >_{\vec{x},e_{\perp}}$$ (6)

¿From the correlation functions, the corresponding correlation lengths can be obtained by integrating and normalizing by the value at r=0:

$$\lambda_B = \int_{-\infty}^{\infty} dr \frac{w\left(r\right)}{w\left(0\right)}$$ (7)

$$\lambda_{RM} = \int_{-\infty}^{\infty} dr \frac{C_{RM}\left(r\right)}{C_{RM}\left(0\right)}$$ (8)

It can be proven that the Fourier transform of the magnetic autocorrelation function is related to the magnetic energy spectrum by

$$\varepsilon_B\left(k\right) = \frac{k^2 w\left(k\right)}{2\left(2\pi\right)^3}$$ (9)

(where $w(\vec{k}) = w(k)$ due to the assumed isotropy) and to the Fourier transform of the autocorrelation function by

$$C_{RM}\left(\vec{k}_\perp\right) = \frac{1}{2}w\left(\vec{k}_\perp, 0\right) \tag{10}$$

With these considerations, the magnetic energy spectrum can be found from an autocorrelation analysis of the rotation map. Furthermore, it was shown that in Fourier space the correlation lengths can be written as

$$\lambda_{RM} = 2\frac{\int_0^\infty w\left(k\right)\,dk}{\int_0^\infty k\,w\left(k\right)\,dk} \tag{11}$$

$$\lambda_B = \pi\frac{\int_0^\infty w(k)\,dk}{\int_0^\infty k^2\,w(k)\,dk} \tag{12}$$

This implies that the RM autocorrelation length has a larger weight on large-scale fluctuations than the magnetic correlation length, and therefore it can be expected that for a broadly peaked magnetic spectrum $\lambda_{RM} > \lambda_B$.

The Fourier space analysis as described above was applied to observational data from Hydra A, Abell 400 and Abell 2634 yielding RM correlation lengths of 2.0 kpc, 5.3 kpc and 8.0 kpc respectively and magnetic field correlation lengths of 0.5 kpc, 2.3 kpc and 4.0 kpc respectively, thus the magnetic field correlation lengths are seen to be indeed smaller than the RM correlation lengths. The RM scalar correlation functions obtained are plotted in Figure 2.

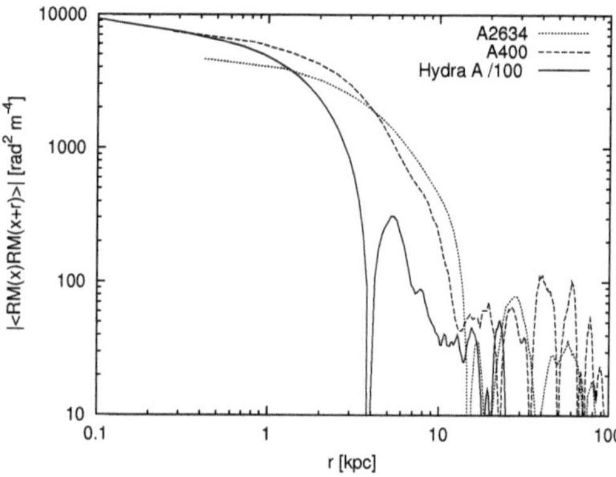

Figure 2: RM autocorrelation functions computed by Vogt and Ensslin from observational data

The calculated magnetic energy spectra (Figure 3)were suppressed at lower values of the wave vector k by the limited window size. The increasing energy density at large k is thought to be due to uncorrelated noise on scales smaller than the beamsize (distance between two adjacent measurements in the RM map). Combining these two effects, the reliable part of the computed spectrum encompasses roughly one order of magnitude in k-space.

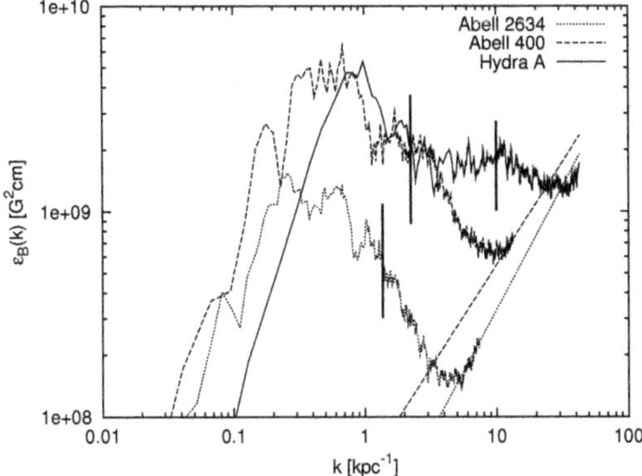

Figure 3: Magnetic energy spectra computed by Vogt and Ensslin from observational data. The Fourier transformed beam sizes are represented by thick vertical lines. The other straight lines correspond to the increase in the energy density for the largest k

In the second stage of developing the method described above, Vogt and Enßlin [5] describe a maximum likelihood analysis of Faraday rotation measures which enables them to conclude a Kolmogorov-like power spectrum of the magnetic energy over at least one order of magnitude in k-space.

1.4 Simulations of Galaxy Clusters Involving Magnetic Fields

Several numerical simulations have been employed in order to model the evolution of magnetic fields with cosmic time and probe the theories postulating different origins of the fields, as well as determine the mechanisms by which the fields are amplified to the present value.

In 1997, Kulsrud et. al. [7] employed the Biermann battery model in a hydrodynamic simulation including the magnetic fields passively (that is, neglecting dynamical importance of the magnetic fields for the gas) and found that in the *protogalactic* era the fields can be amplified from 0 to 10^{-21} G. Following the phase where the Biermann battery appears as an important generator of fields, the simulation by Kulsrud et. al. suggests further amplification of the fields by isotropic Kolmogorov turbulence associated with gravitational structure formation. However, due to the low resolution, the small turbulent scales could not be resolved therefore the computations were done using a simplified analytical model. Results of analytical calculation suggest a significant growth of the magnetic fields, therefore no amplification mechanisms in the galactic era are needed in addition to the protogalactic processes in order to explain the strength measured in the present clusters. Questions nevertheless remained open as to the validity of the analytical model employed by the authors.

Further numerical simulations performed by Banerjee and Jedamzik in 2003 [8] suggested a *primordial* origin of cluster magnetic fields. The authors performed simulations of incompressible, freely decaying viscous magnetohydrodynamics (MHD) in a non-expanding background using 128^3 and 512^3 grids. It was found that fields of up to 10^{-11} G can survive an early universe magnetogenesis (at z=1000) , and that with the decreasing temperature in the universe helical fields tend to arrange such that their correlation length increases.

The most extensive numerical study of cluster magnetic fields was carried out by Dolag et. al. [9] [10]. The authors assumed fields of nanogauss range at a redshift of 15, which are justified by the results of Banerjee (10^{-11} G at redshift 1000 can be transformed to nG order through the expansion of space as prescribed by the Friedmann-Robertson-Walker metric.) A cosmological MHD code was used to evolve the gas from redshift 15 to present in the approximation of smoothed particle hydrodynamics (SPH). Various initial conditions for the magnetic field were analysed in the simulations, namely different average strengths (0.2, 1, and 5 nG) and different distributions (homogeneous and chaotic). It was found that the end-result does not depend on the type of distribution, but only on the average strength of the magnetic seed field. In [9], the average fields at redshift 0 were found to range between 0.4-2.5 μG, thus amplified by a factor of roughly 1000 with respect to the initial seed fields. The computed μG range corresponds to measurements from observational data. Furthermore, the correlation between the X-ray flux and the artificial rotation measure dispersion was established in the form:

$$\sigma_{RM} = A \left(\frac{S_x}{10^{-5} \text{erg/cm}^2/\text{s}} \right)^\alpha \tag{13}$$

where A is a proportionality constant depending on the average temperature of the cluster while α is a power index found from the simulations to be roughly 0.9, varying only slightly between the different simulation runs. From observational data, the index α was computed between 0.8 and 1.15 for various clusters, in agreement with the results of the numerical model. The clear correlation between the X-ray flux and the RM dispersion obtained from the numerical models suggests that the magnetic field is directly related to the gas density.

In 2002, Dolag, Bartelmann and Lesch [10] used the same simulation technique to further investigate the magnetic field structure. They postulate the amplification of magnetic fields by shear flows. A smaller amplification relative to the already very high field strengths at the cluster center is found, thus the field strength profiles flatten off towards small centric distances. In the outer parts of the cluster, the average magnetic field strength is found to be more strongly amplified than what would be expected from simple compression (in which case we expect $B \propto \rho^{2/3}$), which constitutes the effect of shear flows. In the same study, correlation lengths of the magnetic field were computed using an autocorrelation function similar to that described in [6]. Typical values were found between 46.1 and 62.6 kpc, roughly one order of magnitude larger than those discusses by Enßlin and Vogt [6] from observation data. The power spectra computed for the magnetic field followed a linear profile from which the power spectrum was determined between -2.3 and -3.1, much lower than expected from Kolmogorov turbulence (-5/3). The artificial rotation maps however seemed to correlate very well with current observations. One other important fact noted by Dolag et.al. is the large influence of cluster mergers which contribute significantly to increasing the cluster magnetic fields.

The approximation of smoothed particle hydrodynamics is however believed not to be well-suited for resolving shocks and therefore cannot accurately describe cluster mergers. The purpose of the current study is thus to use an adaptive mesh-refinement (AMR) simulation, whose performance in terms of resolving shocks and cluster mergers is improved, and to compare the results with those from observational data and previous numerical models.

2 Theoretical Considerations

As described by the Friedman-Robertson-Walker metric, the universe is expanding, and the rate of expansion is variable in time. In cosmology the concept of "comoving coordinates" has been introduced in order to account for this expansion. The real (spatial) coordinates are related to the comoving coordinates by

$$\vec{r} = a\left(t\right) \vec{x} \tag{14}$$

The time-derivative of $\vec{x}$ is called "peculiar velocity" and is not due to cosmic expansion but to the interaction between individual astrophysical objects. The Hubble constant is defined from observations as $v/r = H$ where v is the receding speed of a galaxy at distance r from the Earth. Taking the derivative of 14 and neglecting peculiar velocities with respect to cosmic expansion rates we find $\vec{v} = d\vec{r}/dt = a\dot{\vec{x}} + \dot{a}\vec{x} \approx \dot{a}\vec{x}$, thus the Hubble parameter in dependence of time reads.

$$H\left(t\right) = \frac{v}{r} = \frac{\dot{a}x}{ax} = \frac{\dot{a}}{a} \tag{15}$$

2.1 Friedman equation

The Friedman equation is the first of six equations which we will employ in order to solve for the cosmic evolution of the galaxy cluster. It relates the time-dependent Hubble parameter to density distributions in the universe:

$$H(t)^2 = \left(\frac{\dot{a}}{a}\right)^2 = H_0^2 \left(\frac{\Omega_r}{a^4} + \frac{\Omega_m}{a^3} + \Omega_\Lambda - \frac{\Omega_c}{a^2}\right) \tag{16}$$

where Ω_m, Ω_r, Ω_Λ, and Ω_c are the present-day densities, respectively, of matter, radiation, cosmological constant and curvature divided by the critical total density which would result in a flat Universe.

2.2 Comoving coordinate transformation

The other five equations needed are the Poisson equation and the four MHD equations (continuity equation, momentum conservation, energy conservation and magnetic field evolution). In order to use the above named equations, we must first transform them to comoving coordinates. For this, we define comoving quantities in terms of the real quantities scaled by a conveniently chosen power of the scale factor a. A summary of scaled comoving quantities as a function of the real-space quantities (denoted with the upper index $*$) is given in 18 below.

$$\rho^* = \frac{\rho}{a^3} \qquad \rho^* \varepsilon^* = \frac{\rho \varepsilon}{a} \tag{17}$$

$$\vec{v}^* = \dot{a}\vec{x} + a\vec{v} \qquad p^* = \frac{p}{a}$$

$$B^* = \frac{B}{\sqrt{a}} \qquad \phi^* = a^2 \phi - \frac{1}{2} a\ddot{a}\vec{x}^2$$

Here, ρ represents the gas density, $\vec{v}$ the velocity, p the pressure, $\vec{B}$ the magnetic field vector, and ε the internal energy defined as

$$\varepsilon = E - \frac{1}{2}|\vec{v}|^2 \tag{18}$$

The scaling for the magnetic field was chosen such that the magnetic pressure, proportional to the square of the field strength, scales with the same power of a as the gas density denoted here as p.

Furthermore, the derivatives with respect to $\vec{r}$ should be expressed in terms of derivatives with respect to $\vec{x}$. The corresponding formulae are given as:

$$\nabla_{\vec{r}} = \frac{1}{a}\nabla_{\vec{x}} \tag{19}$$

$$\left.\frac{\partial}{\partial t}\right|_{\vec{r}=ct} = \left.\frac{\partial}{\partial t}\right|_{\vec{x}=ct} - \frac{\dot{a}}{a}\left(\vec{x} \cdot \nabla_{\vec{x}}\right)$$

2.3 Poisson equation

In physical coordinates, the Poisson equation reads:

$$\nabla^2 \phi^* = 4\pi G \rho^* \tag{20}$$

To transform this equation to comoving coordinates, we use the corresponding scalings for the density ρ, field ϕ and nabla operator ∇ given in the previous section. It follows,

$$\frac{1}{a^2} \nabla \left(a^2 \phi - \frac{1}{2} a \ddot{a} \vec{x}^2 \right) = 4\pi G \frac{\rho}{a^3}$$

$$\nabla \phi - 3\frac{\ddot{a}}{a} = 4\pi G \frac{\rho}{a^3}$$

or, in terms of the comoving mean matter density $\bar{\rho}$,

$$\nabla \phi = 4\pi G \frac{\rho - \bar{\rho}}{a^3} \tag{21}$$

2.4 General considerations of magnetohydrodynamics

Magnetohydrodynamic (MHD) equations describe, as the name suggests, the dynamics of electrically conducting fluids in the presence of magnetic fields. An emphasis is placed on magnetic rather than electric fields although they are, according to electromagnetic field theory, practically equivalent; the reason for this is that free electric charges in the Universe can move to balance out any large-scale electric fields, while magnetic monopoles are believed not to be present in the observable cosmos. This allows for the existence of magnetic fields in extended regions in space, which can become important in astrophysical phenomena. The MHD equations are based on Maxwell's electromagnetic theory, employing several approximations discussed in the following.

Firstly, slow time-variations are assumed such that in Maxwell's equation

$$\nabla \times \vec{B} = \frac{4\pi}{c} \vec{j} + \frac{1}{c} \frac{\partial \vec{E}}{\partial t} \tag{22}$$

the term $c^{-1} \partial \vec{E} / \partial t$ is negligible. This is equivalent to considering electric currents as the only source for magnetic fields.

Secondly, we write the equation of motion for the electron gas in the frame of reference of the positive ions (denoted by the prime) according to [14] as:

$$m_e \frac{d\vec{v}_e{}'}{dt} = -e \left(\vec{E}' + \frac{\vec{v}_e{}'}{c} \times \vec{B}' \right) - m_e \nu_c \vec{v}_e{}' + \text{gravitational and inertia terms} \tag{23}$$

where ν_c is the mean collision frequency associated with momentum transfer ("drag"). Equation (23) assumes that the gas is fully ionized (else, a mean drag force exerted by neutrals on electrons should be included). Furthermore, due to the low electron rest mass, the inertia of the electron is neglected. This means that the electron quickly reaches a balance between the drag and electromagnetic forces, therefore the left-hand side of equation (23) as well as the gravitational and inertia terms can be dropped. Moreover the electron gyromotion is fast compared to macroscopic bulk motions, therefore it does not contribute on average to the conduction current, which entitles us to drop the Lorentz force term. We note that this approximation can be important in regions of strong magnetic field gradients where it becomes inaccurate. With these considerations, equation (23) reduces to a very well known form, namely Ohm's law:

$$\vec{v}_e{}' = -e\vec{E}'/m_e\nu_c \tag{24}$$

$$\vec{j}_e{}' = -en_e\vec{v}_e{}' = \sigma\vec{E}' \tag{25}$$

The third approximation is that collisions between the different components of the plasma occur often enough that the relative velocity between electrons and ions is negligible compared to their actual velocities in the lab frame. Thus, after using the relative velocity to compute the conduction current as described above (here, it cannot be neglected), the fluid is treated as a single-component plasma with a mean fluid velocity, $\vec{u}$, describing all components.

The fourth and final approximation of MHD is that of non-relativistic flows, $u << c$, such that the Lorentz transformations of the electric and magnetic fields from the ion frame of reference to the lab frame can be written as:

$$\vec{B}' = \vec{B} \tag{26}$$

$$\vec{E}' = \vec{E} + \frac{\vec{u}}{c} \times \vec{B} \tag{27}$$

Using (25) and (27) and expressing the conduction current according to (22) neglecting $c^{-1}\partial\vec{E}/\partial t$ as discussed in the first approximation, we obtain

$$\vec{E} = \frac{c}{4\pi\sigma}\nabla \times \vec{B} - \frac{\vec{u}}{c} \times \vec{B}$$

Using Maxwell's equation

$$\nabla \times \vec{E} = -\frac{1}{c}\frac{\partial\vec{B}}{\partial t}$$

we obtain the equation for the time evolution of the magnetic field

$$\frac{\partial\vec{B}}{\partial t} + \nabla \times \left(\vec{B} \times \vec{u}\right) = -\nabla \times \left(\eta\nabla \times \vec{B}\right) \tag{28}$$

where $\eta = c^2/4\pi\sigma$ is the electrical resistivity. Assuming charge neutrality, the influence of the Lorentz force on the dynamics of the fluid can also be calculated using the considerations given above. Here we will not discuss the details of this since the dynamical importance of the magnetic fields was not included in the code of the current simulation.

One important remark remains to be made about equation (28). If the electrical resistivity η can be considered essentially zero we obtain

$$\frac{\partial \vec{B}}{\partial t} + \nabla \times \left(\vec{B} \times \vec{u} \right) = 0$$

This results in the conservation of the magnetic flux Φ through any area element A that moves with the magnetized fluid:

$$\frac{d}{dt} \int_A \vec{B} \cdot d\vec{A} = 0$$

This effect, known as "field freezing", is two-fold: free charged particles are well tied to the magnetic lines of force and simultaneously the field is tied to the fluid such that the flux through any area moving with the conducting plasma is conserved. Field freezing has many applications in astrophysics not because of unusually high plasma conductivities but because the right-hand side of equation (28) represents a diffusive term with a characteristic diffusion time $t_D \approx L^2/\eta$ where L is the length scale of the field; on large astrophysical scales t_D is very large which makes the effect of diffusion negligible. In the current simulation the field freezing approximation was also involved.

2.5 MHD continuity equation

After an introduction to the general considerations of the MHD equations, we continue by transforming the MHD equations to comoving coordinates. The first equation we consider is the continuity equation which in real coordinates reads:

$$\left. \frac{\partial \rho^*}{\partial t} \right|_{\vec{r}=ct} + \nabla_{\vec{r}} \left(\rho^* \vec{v}^* \right) = 0 \tag{29}$$

Using the scaling conventions named in equations 18 and 20 above, it follows:

$$\frac{\partial}{\partial t} \left(\frac{\rho}{a^3} \right) - \frac{\dot{a}}{a} \left(\vec{x} \cdot \nabla \frac{\rho}{a^3} \right) + \frac{1}{a} \nabla \left[\frac{\rho}{a^3} \left(\dot{a}\vec{x} + a\vec{v} \right) \right] = 0$$

Using that $\nabla \vec{x} = 3$ and that a does not depend on $\vec{x}$

$$\Rightarrow \frac{1}{a^3} \frac{\partial \rho}{\partial t} + \rho(-3) \frac{\dot{a}}{a^4} - \frac{\dot{a}}{a^4} \vec{x} \cdot \nabla \rho + \frac{\dot{a}}{a^4} \nabla \rho \cdot \vec{x} + \frac{1}{a^3} \nabla \rho \cdot \vec{v} + \frac{\rho}{a^4} \left(3\dot{a} + a\nabla \vec{v} \right) = 0$$

$$\Rightarrow \frac{1}{a^3} \frac{\partial \rho}{\partial t} + \frac{1}{a^3} \nabla \rho \vec{v} + \frac{1}{a^3} \rho \nabla \vec{v} = 0$$

Our final equation in comoving coordinates is then written as

$$\frac{\partial \rho}{\partial t} + \nabla \left(\rho \vec{v} \right) = 0 \tag{30}$$

2.6 MHD momentum conservation

The equation of conservation of momentum in real coordinates assuming an inviscid fluid is

$$\frac{\partial \rho^* \vec{v}^*}{\partial t} + \nabla \left(\rho^* \vec{v}^* \vec{v}^* - \vec{B}^* \vec{B}^* \right) + \nabla p_{tot}^* + \rho^* \nabla \phi^* = 0 \tag{31}$$

where $\vec{v}^* \vec{v}^*$ is the dyadic tensor formed by $\vec{v}_i \vec{v}_j$ and p_{tot}^* is the total pressure: $p_{tot}^* = p_{static}^* + \frac{B^{*\,2}}{2}$. Transforming to comoving coordinates gives:

$$\frac{\partial}{\partial t} \left(\frac{\rho}{a^3} \left(\dot{a}\vec{x} + a\vec{v} \right) \right) - \frac{\dot{a}}{a} \left(\vec{x} \cdot \nabla \right) \left[\frac{\rho}{a^3} \left(\dot{a}\vec{x} + a\vec{v} \right) \right] + \frac{1}{a^4} \nabla_i \cdot \left[\rho \left(\dot{a}x_i + av_i \right) \left(\dot{a}x_j + av_j \right) \right] -$$

$$- \frac{\nabla \vec{B}\vec{B}}{a^2} + \frac{\nabla p}{a^2} + \frac{\rho}{a^4} \nabla \left(a^2 \phi - \frac{1}{2} a\ddot{a}\vec{x}^2 \right) = 0$$

Further note that $\left. \frac{\partial \vec{x}}{\partial t} \right|_{\vec{x}=ct} = 0$. For the first term in the previous equation we obtain:

$$\frac{\partial}{\partial t} \left(\frac{\rho}{a^3} \left(\dot{a}\vec{x} + a\vec{v} \right) \right) = -3a^{-4}\dot{a}\rho \left(\dot{a}\vec{x} + a\vec{v} \right) + a^{-3}\dot{\rho} \left(\dot{a}\vec{x} + a\vec{v} \right) + a^{-3}\rho \left(\ddot{a}\vec{x} + \dot{a}\vec{v} + a\dot{\vec{v}} \right)$$

The second term we express as:

$$\frac{\dot{a}}{a} \left(\vec{x} \cdot \nabla \right) \left[\frac{\rho}{a^3} \left(\dot{a}\vec{x} + a\vec{v} \right) \right] = \dot{a}^2 a^{-4} \vec{x} \cdot \nabla \left(\rho \vec{x} \right) + \dot{a} a^{-3} \vec{x} \cdot \nabla \left(\rho \vec{v} \right)$$

where $\vec{x} \cdot \nabla \left(\rho \vec{x} \right) = \vec{x} \left(\vec{x} \cdot \nabla \rho \right) + \rho \vec{x}$ and $\vec{x} \cdot \nabla \left(\rho \vec{v} \right) = \vec{v} \left(\vec{x} \cdot \nabla \rho \right) + \rho \left(\vec{x} \cdot \nabla \right) \vec{v}$. Moving on to the third term, the divergence of the dyadic tensor, we obtain:

$$\frac{1}{a^4} \nabla_i \cdot \left[\rho \left(\dot{a}x_i + av_i \right) \left(\dot{a}x_j + av_j \right) \right] = a^{-4} \nabla \rho \left[\dot{a}^2 x_i x_j + a^2 v_i v_j + a\dot{a}x_i v_j + a\dot{a}x_j v_i \right] +$$

$$+ a^{-4} \rho \left[\dot{a}^2 \nabla_i x_i x_j + a\dot{a}\nabla_i x_j v_i + a\dot{a}\nabla_i x_i v_j + a^2 \nabla_i v_i v_j \right]$$

With

$$\nabla \rho \left(x_i x_j \right) = \frac{\partial \rho}{\partial x_i} x_i x_j = \vec{x} \left(\nabla \rho \cdot \vec{x} \right)$$

$$\nabla \rho \left(v_i x_j \right) = \frac{\partial \rho}{\partial x_i} v_i x_j = \vec{x} \left(\nabla \rho \cdot \vec{v} \right)$$

$$\nabla \rho \left(x_i v_j \right) = \frac{\partial \rho}{\partial x_i} x_i v_j = \vec{v} \left(\nabla \rho \cdot \vec{x} \right)$$

$$\nabla \rho \left(v_i v_j \right) = \frac{\partial \rho}{\partial x_i} v_i v_j = \vec{v} \left(\nabla \rho \cdot \vec{v} \right)$$

$$\nabla_i \left(x_i x_j \right) = \frac{\partial x_i}{\partial x_i} x_j + x_i \frac{\partial x_j}{\partial x_i} = 4\vec{x}$$

$$\nabla_i x_i v_j + \nabla_i x_j v_i = \frac{\partial x_i}{\partial x_i} v_j + \frac{\partial x_j}{\partial x_i} v_i + \frac{\partial v_i}{\partial v_j} x_j + \frac{\partial x_i}{\partial x_i} x_i = 4\vec{v} + \left(\vec{x} \cdot \nabla \right) \vec{v} + \vec{x} \left(\nabla \cdot \vec{v} \right)$$

The fourth and fifth terms need no further simplification. The last term can be written as

$$\frac{\rho}{a^4}\nabla\left(a^2\phi-\frac{1}{2}a\ddot{a}\vec{x}^2\right)=\frac{1}{a^2}\rho\nabla\phi-\frac{\ddot{a}}{a^3}\rho\vec{x}$$

Putting everything together,

$$-3a^{-4}\dot{a}^2\rho\vec{x}-3a^{-3}\dot{a}\rho\vec{v}+a^{-3}\dot{a}\rho\vec{x}+a^{-2}\dot{\rho}\vec{v}+a^{-3}\ddot{a}\rho\vec{x}+a^{-2}\rho\dot{\vec{v}}+a^{-3}\dot{a}\rho\vec{v}-$$

$$a^{-4}\dot{a}^2\vec{x}\left(\nabla\rho\cdot\vec{x}\right)-a^{-4}\dot{a}^2\rho\vec{x}-a^{-3}\dot{a}\left(\vec{x}\cdot\nabla\rho\right)\vec{v}-a^{-3}\dot{a}\rho\left(\vec{x}\cdot\nabla\right)\vec{v}$$

$$+a^{-4}\dot{a}^2\vec{x}\left(\nabla\rho\cdot\vec{x}\right)+a^{-3}\dot{a}\vec{x}\left(\vec{v}\cdot\nabla\rho\right)+a^{-3}\dot{a}\vec{v}\left(\nabla\rho\cdot\vec{x}\right)+a^{-2}\nabla\rho v_i v_j$$

$$+4a^{-4}\dot{a}^2\rho\vec{x}+4a^{-3}\dot{a}\rho\vec{v}+a^{-3}\dot{a}\rho\left(\vec{x}\cdot\nabla\right)\vec{v}+a^{-3}\dot{a}\rho\vec{x}\left(\nabla\cdot\vec{v}\right)+a^{-2}\rho\nabla_i v_i v_j$$

$$-a^{-2}\nabla\vec{B}\vec{B}+a^{-2}\nabla p+a^{-2}\rho\nabla\phi-a^{-3}\ddot{a}\rho\vec{x}=0$$

Multiplying both sides by a^2 and cancelling terms where possible, we obtain:

$$\frac{\dot{a}}{a}\rho\vec{x}+\dot{\rho}\vec{v}+\rho\dot{\vec{v}}+2\frac{\dot{a}}{a}\rho\vec{v}+\frac{\dot{a}}{a}\vec{x}\left(\vec{v}\cdot\nabla\rho\right)+\frac{\dot{a}}{a}\vec{x}\rho\left(\nabla\cdot\vec{v}\right)+\rho\nabla_i v_i v_j+\nabla\rho v_i v_j-\nabla\vec{B}\vec{B}+\rho\nabla\phi+\nabla p=0$$

Combining the terms

$$\dot{\rho}\vec{v}+\rho\dot{\vec{v}}\;=\;\frac{\partial\rho\vec{v}}{\partial t}$$

$$\rho\nabla_i v_i v_j+\nabla\rho v_i v_j\;=\;\nabla\rho\vec{v}\vec{v}$$

$$\vec{x}\left(\vec{v}\cdot\nabla\rho\right)+\vec{x}\rho\left(\nabla\cdot\vec{v}\right)\;=\;\vec{x}\nabla\cdot\left(\rho\vec{v}\right)$$

and grouping the terms containing $\vec{x}$ we obtain:

$$\frac{\partial\rho\vec{v}}{\partial t}+\nabla\left(\rho\vec{v}\vec{v}-\vec{B}\vec{B}\right)+2\frac{\dot{a}}{a}\rho\vec{v}+\rho\nabla\phi+\nabla p+\frac{\dot{a}}{a}\vec{x}\left[\dot{\rho}+\nabla\cdot\left(\rho\vec{v}\right)\right]=0$$

Using the continuity equation derived earlier, $\dot{\rho}+\nabla\cdot\left(\rho\vec{v}\right)=0$ thus we obtain as our final result:

$$\frac{\partial\rho\vec{v}}{\partial t}+\nabla\left(\rho\vec{v}\vec{v}-\vec{B}\vec{B}\right)+2\frac{\dot{a}}{a}\rho\vec{v}+\rho\nabla\phi+\nabla p=0 \tag{32}$$

2.7   Energy conservation

The MHD equation expressing the conservation of energy reads, in physical coordinates:

$$\frac{\partial\rho^*E^*}{\partial t}+\nabla\cdot\left(\vec{v}^*\left(\rho^*E^*+p^*_{tot}\right)-\vec{B}^*\left(\vec{v}^*\cdot\vec{B}^*\right)\right)+\rho^*\nabla\phi^*\cdot\vec{v}^*=0 \tag{33}$$

In cases where the kinetic energy is dominant it is however preferred for computing purposes to use the equation for the conservation of the internal energy, related to the total energy according to 18. This equation reads (here: disregarding contributions from the magnetic field)

$$\frac{\partial\rho^*\varepsilon^*}{\partial t}+\nabla\cdot\left[\left(\rho^*\varepsilon^*+p^*\right)\vec{v}^*\right]-\vec{v}^*\cdot\nabla p^*=0 \tag{34}$$

We obtain by comoving coordinate transformation:

$$\frac{\partial}{\partial t}\left(\frac{\rho\varepsilon}{a}\right) - \frac{\dot{a}}{a}\left(\vec{x}\cdot\nabla\frac{\rho\varepsilon}{a}\right) + \frac{1}{a}\nabla\left[\left(\frac{\rho\varepsilon}{a}+\frac{p}{a}\right)(\dot{a}\vec{x}+a\vec{v})\right] - \frac{1}{a}(\dot{a}\vec{x}+a\vec{v})\nabla\frac{p}{a} = 0$$

$$\Rightarrow \frac{1}{a}\frac{\partial\rho\varepsilon}{\partial t} + \rho\varepsilon\left(-\frac{1}{a^2}\right)\dot{a} - \frac{\dot{a}}{a^2}\vec{x}\cdot\nabla\rho\varepsilon + \frac{1}{a^2}\left[\nabla(\rho\varepsilon+p)\right](\dot{a}\vec{x}+a\vec{v}) +$$

$$\frac{1}{a^2}(\rho\varepsilon+p)(3\dot{a}+a\nabla\vec{v}) - \frac{\dot{a}\vec{x}+a\vec{v}}{a^2}\nabla p = 0$$

$$\Rightarrow \frac{1}{a}\frac{\partial\rho\varepsilon}{\partial t} - \frac{\dot{a}}{a}\rho\varepsilon - \frac{\dot{a}}{a^2}\vec{x}\cdot\nabla\rho\varepsilon + \frac{1}{a^2}(\nabla\rho\varepsilon)\dot{a}\vec{x} + \frac{1}{a^2}\nabla\rho\varepsilon a\vec{v} + \frac{1}{a^2}\nabla p\dot{a}\vec{x} + \frac{1}{a^2}\nabla pa\vec{v} + \frac{3\rho\varepsilon\dot{a}}{a^2} +$$

$$+\frac{3p\dot{a}}{a^2} + \frac{\rho\varepsilon}{a^2}a\nabla\vec{v} + \frac{p}{a^2}a\nabla\vec{v} - \frac{\dot{a}}{a^2}\vec{x}\nabla p - \frac{\vec{v}}{a}\nabla p = 0$$

Canceling out the terms where possible and combining terms of the form $\vec{v}\cdot\nabla p + p\nabla\vec{v}$ into $\nabla p\vec{v}$ and $\vec{v}\cdot\nabla\rho\varepsilon + \rho\varepsilon\nabla\vec{v}$ into $\nabla\rho\varepsilon\vec{v}$ we finally obtain after multiplying both sides of the equation by a:

$$\frac{\partial\rho\varepsilon}{\partial t} + \nabla\cdot\left[(\rho\varepsilon+p)\vec{v}\right] - \vec{v}\nabla p + \frac{\dot{a}}{a}(2\rho\varepsilon+3p) = 0$$

Using that $p = \rho\varepsilon(\gamma-1)$ it follows

$$\frac{\partial\rho\varepsilon}{\partial t} + \nabla\cdot\left[(\rho\varepsilon+p)\vec{v}\right] - \vec{v}\nabla p + \frac{\dot{a}}{a}\rho\varepsilon(3\gamma-1) = 0 \qquad (35)$$

2.8 MHD magnetic field evolution

To describe the evolution of the magnetic field we use in real coordinates the following equation (assuming zero resistivity of the fluid):

$$\frac{\partial\vec{B}^*}{\partial t} + \nabla\cdot\left(\vec{v}^*\vec{B}^* - \vec{B}^*\vec{v}^*\right) = \frac{\nabla p^* \times \nabla\rho^*}{\rho^{*2}}\frac{cm_H}{e}\frac{1}{1+\chi} \qquad (36)$$

where the RHS of the equation contains the Biermann battery effect (generation of magnetic field when ∇p and $\nabla\rho$ are not parallel). χ denotes the ionization factor and is assumed to be constant in space. We obtain:

$$\frac{\partial}{\partial t}\frac{\vec{B}}{\sqrt{a}} - \frac{\dot{a}}{a}(\vec{x}\cdot\nabla)\frac{\vec{B}}{\sqrt{a}} + \frac{1}{a\sqrt{a}}\nabla_i\cdot\left[(av_i+\dot{a}x_i)B_j - B_i(av_j+\dot{a}x_j)\right] = \frac{cm_H}{e}\frac{1}{1+\chi}\frac{\frac{\nabla p}{a^2}\times\frac{\nabla\rho}{a^4}}{\frac{\rho^2}{a^6}}$$

$$\Rightarrow \frac{1}{\sqrt{a}}\frac{\partial\vec{B}}{\partial t} - \frac{\dot{a}}{2a\sqrt{a}}\vec{B} - \frac{\dot{a}}{a\sqrt{a}}(\vec{x}\cdot\nabla)\vec{B} + \frac{1}{a\sqrt{a}}\nabla_i\cdot[av_iB_j + \dot{a}x_iB_j - B_iav_j + B_i\dot{a}x_j] =$$

$$= \frac{cm_H}{e}\frac{1}{1+\chi}\frac{\nabla p\times\nabla\rho}{\rho^2}$$

Let us look closer at the term $\nabla_i\cdot[av_iB_j + \dot{a}x_iB_j - B_iav_j + B_i\dot{a}x_j]$. This can be written as

$$a\nabla_i\cdot(v_iB_j - B_iv_j) + \dot{a}\nabla_i\cdot(x_iB_j - B_ix_j) =$$

$$= a\nabla \cdot \left(\vec{v}\vec{B} - \vec{B}\vec{v}\right) + \dot{a}\left[\frac{\partial x_i}{x_i}B_j + x_i\frac{\partial B_j}{x_i} - \frac{\partial x_j}{x_i}B_i - x_j\frac{\partial B_i}{x_i}\right]$$

$$= a\nabla \cdot \left(\vec{v}\vec{B} - \vec{B}\vec{v}\right) + \dot{a}\left[3\vec{B} + (\vec{x}\cdot\nabla)\vec{B} - \vec{B} - \vec{x}\nabla\cdot\vec{B}\right]$$

We know furthermore that $\nabla \cdot \vec{B} = 0$ from Maxwell's equations. Going back to the original equation with these in mind, we obtain:

$$\frac{1}{\sqrt{a}}\frac{\partial \vec{B}}{\partial t} - \frac{\dot{a}}{2a\sqrt{a}}\vec{B} - \frac{\dot{a}}{a\sqrt{a}}(\vec{x}\cdot\nabla)\vec{B} + \frac{\dot{a}}{a\sqrt{a}}\left[2\vec{B} + (\vec{x}\cdot\nabla)\vec{B}\right] + \frac{1}{\sqrt{a}}\nabla\cdot\left(\vec{v}\vec{B} - \vec{B}\vec{v}\right)$$

$$= \frac{cm_H}{e}\frac{1}{1+\chi}\frac{\nabla p \times \nabla \rho}{\rho^2}$$

Simplifying and multiplying both sides by $\sqrt{a}$, we arrive at:

$$\frac{\partial \vec{B}}{\partial t} + \nabla\cdot\left(\vec{v}\vec{B} - \vec{B}\vec{v}\right) + \frac{3}{2}\frac{\dot{a}}{a}\vec{B} = \sqrt{a}\frac{cm_H}{e}\frac{1}{1+\chi}\frac{\nabla p \times \nabla \rho}{\rho^2} \tag{37}$$

This concludes the set of equations needed in order to describe the cosmic evolution of a galaxy cluster.

3 Methods

3.1 Principle and Initial Conditions

We performed an adaptive mesh refinement hydrodynamic simulation using the Flash2.4 code [12]. We worked in Cartesian geometry; the simulated region was a cube of side length 64 Mpc. Initial conditions inside this region were chosen according to the Santa Barbara Cluster project [11]. The initial gas density was thus governed by random Gauss fluctuations of asymptotic spectral index n=1. In addition to the gas (assumed to be ideal), 2.09 million massive dark matter particles were involved in the simulation. In addition, we chose a uniform initial magnetic field of 10^{-9} G in the (1,1,1) direction, based on the findings by Dolag et.al. [9] which revealed that both uniform and chaotic seed fields produce the same result.

¿From these initial conditions, the system was evolved from redshift 50 to redshift 0 by numerically solving equations (16), (21), (30), (32), (35) and (37). The magnetic field was passively implemented, meaning that the dynamic terms containing the magnetic field were neglected in equations (32) (momentum conservation) and (35) (energy conservation). Furthermore, upon a test run, the Biermann battery term was found to be negligible in equation (37) and was therefore omitted from further simulation runs.

3.2 Cosmology

The Friedman equation, (16), is solved for the scale factor $a(t)$ and its time derivative $\dot{a}(t)$ by the Cosmology module implemented in Flash2.4. Any accurate ordinary differential equation (ODE) solver can be used; in this case, a variable-order Adams algorithm is implemented in Flash. The cosmology was assumed to be Λ-Cold Dark Matter, characterized by $\Omega_m = 0.3$ and $\Omega_\Lambda = 0.7$.

3.3 Gravitational interaction of gas and dark matter

The Poisson equation, (21), for the gas and dark matter is solved using a multigrid Fast Fourier Transform (FFT) algorithm implemented in the Gravity module. Unlike other quantities which are evolved on a grid, the Particle module (which governs the motion of dark-matter particles) follows the motion of Lagrangian mass tracers. The dark-matter-related quantities (position, mass, velocity) are evolved in each time step independent of the grid using the Particle module and the result is then interpolated onto the grid which enables the gravitational interaction of dark matter and gas to be computed and included in the gas dynamics. Periodic boundary conditions are assumed throughout the simulation.

3.4 Adaptive Mesh Refinement

The mesh refinement criterion in the simulation was based on the number of dark matter particles contained in a cell. The grid is divided in blocks, each block containing 16^3 cells. The gas-related quantities (density, pressure, temperature) are assumed to be constant in each cell. When the dark matter density interpolated on the grid reaches a level higher than 4 DM particles per cell in a certain region of the simulation, the block corresponding to that region is split into 8 blocks of side length equal to half the side length of the initial block. This allows for smaller cell sizes, thus better resolution of that particular region, which is a region of interest since an agglomeration of DM particles implies cosmic structure formation. The current simulation had an effective resolution of 1024^3 cells, which means that if the entire simulation volume had been resolved at the highest resolution used in the computation according to mesh refinement, the result of the simulation would comprise 1024^3 data point-arrays for each of the quantities computed.

3.5 Hydrodynamics and the Riemann problem

The gas is not only under the influence of the gravitational potential whose computation has been described in the previous section, but also under the influence of pressure, magnetic field and viscous forces. The system is thus best described by the full MHD equations. However, the complexity of solving these equations is very high therefore some approximations are needed.

Firstly, since we assume the gas to be ideal, we will neglect its viscosity and resistivity. Secondly, since many numerical simulations which did not include the magnetic field were until now successful at producing galaxy clusters which fit the current observations quite well, we infer that the magnetic field does not have a crucial role in structure formation, and therefore we neglect the dynamic terms containing the magnetic field. This of course impedes us from drawing any conclusions as to exactly how much the magnetic field influences galaxy clusters, but enables us to predict how the field evolves due to structure formation, what mechanisms seem to amplify it and what we can expect from its present general properties. Further simplifications in the code are the absence of a cooling mechanism and neglecting the effects of star formation, which would also render the complexity uncomputable. Since the smallest cell in the simulation box is roughly 30 kpc wide, star formation could not be resolved even if it were included in the code.

The MHD equations in the absence of a magnetic field are also called the Euler equations. The Flash2.4 module which solves these Euler equations (also in comoving coordinates) is the Hydrodynamics module which uses a piecewise parabolic method (PPM) to solve the Riemann problem. The Riemann problem consists in computing the resulting gas-related quantities (density, pressure, velocity etc.) following the breakup of a discontinuity, which initially separates two arbitrary constant states, in our case any two adjacent cells. The piecewise parabolic method refers to the fact that the quantities computed at the cell boundaries are interpolated to and from the cell centers using parabolic functions (polynomials of second degree). Riemann solvers are known to accurately evolve shocks, and this may prove important for the evolution of magnetic fields which are believed to grow significantly during cluster mergers [10].

3.6 Evolution of the magnetic field

The magnetic field was evolved separately after the completion of each timestep in the integration of the hydrodynamic equations using:

$$\frac{\partial \vec{B}}{\partial t} - \nabla \times \left(\vec{v} \times \vec{B} \right) + \frac{3}{2} \frac{\dot{a}}{a} \vec{B} = 0 \tag{38}$$

which is the equivalent to equation (37) neglecting the Biermann battery term.

A common problem in solving for the magnetic field is the requirement from Maxwell's equations that the solution should be divergence-free. In our simulation, this was accounted for by evolving the magnetic field on a staggered mesh as described in [13]. The method is based on calculating the magnetic field on the center of the cell faces and the cross product $\vec{v} \times \vec{B}$ on the centers of the cell edges, rather than computing both $\vec{B}$ and $\vec{v} \times \vec{B}$ at the cell center. By these means, the divergence of the magnetic field is automatically conserved when solving the equation

$$\frac{\partial \vec{B}}{\partial t} = \nabla \times \left(\vec{v} \times \vec{B} \right) \tag{39}$$

Note that the equation we want to solve, namely equation (38), contains an additional cosmological term compared to equation (39). This was overcome by analytically solving:

$$\frac{\partial \vec{B}}{\partial t} + \frac{3}{2}\frac{\dot{a}}{a}\vec{B} = 0 \tag{40}$$

by an exponential function which was multiplied after each time-step with the result of evolving equation (39) numerically.

We present below a proof of the claim that evolving the magnetic field on the center of the cell faces and the cross product $\vec{v} \times \vec{B}$ on the centers of the cell edges using equation (39) conserves the divergence of the magnetic field. Let us denote the indices describing the position of the cell center by (i,j,k) where i corresponds to the x direction, j to y and k to z. The neighboring cells are then located at (i+1,j,k), (i-1,j,k), (i, j+1,k) and so on. The centers of cell faces have coordinates (i+1/2,j,k), (i-1/2,j,k), (i,j+1/2,k) etc. The centers of cell edges are located at (i+1/2,j+1/2,k), (i-1/2,j+1/2,k), (i+1/2,j,k+1/2), etc. According to the prescription given in [13], the solution to (39) will be given by:

$$
\begin{aligned}
B^{x,n+1}_{i+1/2,j,k} = B^{x,n}_{i+1/2,j,k} \;+\;& \frac{\Delta t}{\Delta y}\left(\left(\vec{v}\times\vec{B}\right)^{z}_{i+1/2,j+1/2,k} - \left(\vec{v}\times\vec{B}\right)^{z}_{i+1/2,j-1/2,k} \right) \\
-\;& \frac{\Delta t}{\Delta z}\left(\left(\vec{v}\times\vec{B}\right)^{y}_{i+1/2,j,k+1/2} - \left(\vec{v}\times\vec{B}\right)^{y}_{i+1/2,j,k-1/2} \right) \\
B^{y,n+1}_{i,j+1/2,k} = B^{y,n}_{i,j+1/2,k} \;+\;& \frac{\Delta t}{\Delta z}\left(\left(\vec{v}\times\vec{B}\right)^{x}_{i,j+1/2,k+1/2} - \left(\vec{v}\times\vec{B}\right)^{x}_{i,j+1/2,k-1/2} \right) \\
-\;& \frac{\Delta t}{\Delta x}\left(\left(\vec{v}\times\vec{B}\right)^{z}_{i+1/2,j+1/2,k} - \left(\vec{v}\times\vec{B}\right)^{z}_{i-1/2,j+1/2,k} \right) \\
B^{z,n+1}_{i,j,k+1/2} = B^{z,n}_{i,j,k+1/2} \;+\;& \frac{\Delta t}{\Delta x}\left(\left(\vec{v}\times\vec{B}\right)^{y}_{i+1/2,j,k+1/2} - \left(\vec{v}\times\vec{B}\right)^{y}_{i-1/2,j,k+1/2} \right) \\
-\;& \frac{\Delta t}{\Delta y}\left(\left(\vec{v}\times\vec{B}\right)^{x}_{i,j+1/2,k+1/2} - \left(\vec{v}\times\vec{B}\right)^{x}_{i,j-1/2,k+1/2} \right)
\end{aligned}
$$

The divergence after the (n+1)-th iteration step is given by:

$$
\begin{aligned}
\left(\nabla \cdot \vec{B}\right)^{n+1} &= \frac{\partial B^{x,n+1}}{\partial x} + \frac{\partial B^{y,n+1}}{\partial y} + \frac{\partial B^{z,n+1}}{\partial z} \\
&= \frac{B^{x,n+1}_{i+1/2,j,k} - B^{x,n+1}_{i-1/2,j,k}}{\Delta x} + \frac{B^{y,n+1}_{i,j+1/2,k} - B^{y,n+1}_{i,j-1/2,k}}{\Delta y} + \frac{B^{z,n+1}_{i,j,k+1/2} - B^{z,n+1}_{i,j,k-1/2}}{\Delta z}
\end{aligned}
$$

Using the (n+1)-th iteration values of B as a function of the n-th iteration values of B as defined above, the divergence becomes (using the shorthand notation $\vec{v} \times \vec{B} = \vec{X}$):

$$
\begin{aligned}
\left(\nabla \cdot \vec{B}\right)^{n+1} = {}& \frac{1}{\Delta x} B^{x,n}_{i+1/2,j,k} + \frac{\Delta t}{\Delta y \Delta x}\left(\vec{X}^{z}_{i+1/2,j+1/2,k} - \vec{X}^{z}_{i+1/2,j-1/2,k}\right) \\
& - \frac{\Delta t}{\Delta z \Delta x}\left(\vec{X}^{y}_{i+1/2,j,k+1/2} - \vec{X}^{y}_{i+1/2,j,k-1/2}\right) \\
& - \frac{1}{\Delta x} B^{x,n}_{i-1/2,j,k} - \frac{\Delta t}{\Delta y \Delta x}\left(\vec{X}^{z}_{i-1/2,j+1/2,k} - \vec{X}^{z}_{i-1/2,j-1/2,k}\right) \\
& + \frac{\Delta t}{\Delta z \Delta x}\left(\vec{X}^{y}_{i-1/2,j,k+1/2} - \vec{X}^{y}_{i-1/2,j,k-1/2}\right) \\
& + \frac{1}{\Delta y} B^{y,n}_{i,j+1/2,k} + \frac{\Delta t}{\Delta z \Delta y}\left(\vec{X}^{x}_{i,j+1/2,k+1/2} - \vec{X}^{x}_{i,j+1/2,k-1/2}\right) \\
& - \frac{\Delta t}{\Delta y \Delta x}\left(\vec{X}^{z}_{i+1/2,j+1/2,k} - \vec{X}^{z}_{i-1/2,j+1/2,k}\right) \\
& - \frac{1}{\Delta y} B^{y,n}_{i,j-1/2,k} - \frac{\Delta t}{\Delta y \Delta z}\left(\vec{X}^{x}_{i,j-1/2,k+1/2} - \vec{X}^{x}_{i,j-1/2,k-1/2}\right) \\
& + \frac{\Delta t}{\Delta y \Delta x}\left(\vec{X}^{z}_{i+1/2,j-1/2,k} - \vec{X}^{z}_{i-1/2,j-1/2,k}\right) \\
& + \frac{1}{\Delta z} B^{z,n}_{i,j,k+1/2} + \frac{\Delta t}{\Delta z \Delta x}\left(\vec{X}^{y}_{i+1/2,j,k+1/2} - \vec{X}^{y}_{i-1/2,j,k+1/2}\right) \\
& - \frac{\Delta t}{\Delta y \Delta z}\left(\vec{X}^{x}_{i,j+1/2,k+1/2} - \vec{X}^{x}_{i,j-1/2,k+1/2}\right) \\
& - \frac{1}{\Delta z} B^{z,n}_{i,j,k-1/2} + \frac{\Delta t}{\Delta y \Delta z}\left(\vec{X}^{x}_{i,j+1/2,k-1/2} - \vec{X}^{x}_{i,j-1/2,k-1/2}\right) \\
& - \frac{\Delta t}{\Delta z \Delta x}\left(\vec{X}^{y}_{i+1/2,j,k-1/2} - \vec{X}^{y}_{i-1/2,j,k-1/2}\right)
\end{aligned}
$$

Grouping the terms containing $B^{*,n}_{*,*,*}$ we observe that we obtain exactly $(\nabla \cdot B)^n$. Factoring the terms containing $\frac{\Delta t}{\Delta x \Delta y}$, we obtain

$$
\begin{aligned}
\frac{\Delta t}{\Delta x \Delta y}(&\vec{X}^{z}_{i+1/2,j+1/2,k} - \vec{X}^{z}_{i+1/2,j-1/2,k} - \vec{X}^{z}_{i-1/2,j+1/2,k} + \vec{X}^{z}_{i-1/2,j-1/2,k} \\
& - \vec{X}^{z}_{i+1/2,j+1/2,k} + \vec{X}^{z}_{i-1/2,j+1/2,k} + \vec{X}^{z}_{i+1/2,j-1/2,k} - \vec{X}^{z}_{i-1/2,j-1/2,k}) = 0
\end{aligned}
$$

Factoring the terms containing $\frac{\Delta t}{\Delta x \Delta z}$, we obtain in the same fashion

$$
\begin{aligned}
\frac{\Delta t}{\Delta x \Delta z}\Big(& -\vec{X}^{y}_{i+1/2,j,k+1/2} + \vec{X}^{y}_{i+1/2,j,k-1/2} + \vec{X}^{y}_{i-1/2,j,k+1/2} - \vec{X}^{y}_{i-1/2,j,k-1/2} \\
& + \vec{X}^{y}_{i+1/2,j,k+1/2} - \vec{X}^{y}_{i-1/2,j,k+1/2} - \vec{X}^{y}_{i+1/2,j,k-1/2} + \vec{X}^{y}_{i-1/2,j,k-1/2}\Big) = 0
\end{aligned}
$$

Finally, factoring the terms containing $\frac{\Delta t}{\Delta y \Delta z}$, we obtain

$$
\begin{aligned}
\frac{\Delta t}{\Delta y \Delta z}\Big(& \vec{X}^{x}_{i,j+1/2,k+1/2} - \vec{X}^{x}_{i,j+1/2,k-1/2} - \vec{X}^{x}_{i,j-1/2,k+1/2} + \vec{X}^{x}_{i,j-1/2,k-1/2} \\
& - \vec{X}^{x}_{i,j+1/2,k+1/2} + \vec{X}^{x}_{i,j-1/2,k+1/2} + \vec{X}^{x}_{i,j+1/2,k-1/2} - \vec{X}^{x}_{i,j-1/2,k-1/2}\Big) = 0
\end{aligned}
$$

Thus, all the terms containing $\vec{X}$ exactly cancel and we are left with the conserved divergence:

$$(\nabla \cdot B)^{n+1} = (\nabla \cdot B)^n \tag{41}$$

Thus, if the initial seed field is chosen to be divergence-free, the field will stay divergence-free throughout the simulation using this method.

One last aspect which requires discussion regarding this method is how to "translate" the magnetic fields and gas velocities from the cell center to the cell edge or face center. This is done by Lagrange polynomial interpolation using data from several adjacent cell centers. In this simulation, a fifth order Lagrange interpolation was employed.

4 Results

4.1 Structure formation

We started with randomly distributed density perturbations which enabled the regions of slightly higher density to start falling towards each other under the influence of gravity. Eventually, the dark matter accumulates in the central part of the simulation volume forming a gravitational potential well into which the gas falls. This results in an increasing density and temperature of the gas and finally in the formation of a galaxy cluster which can be seen at the intersection of filaments of higher density than the background. The in-fall of the gas into the gravitational well can result into temperatures of up to 10^8 K, which agree with and explain the observational data. A time-evolution sequence of the gas density and temperature is plotted below for four different times. The figures represent sections in the (x,y) plane through the middle of the simulation box. The correlation between increasing temperature and structure formation can clearly be seen.

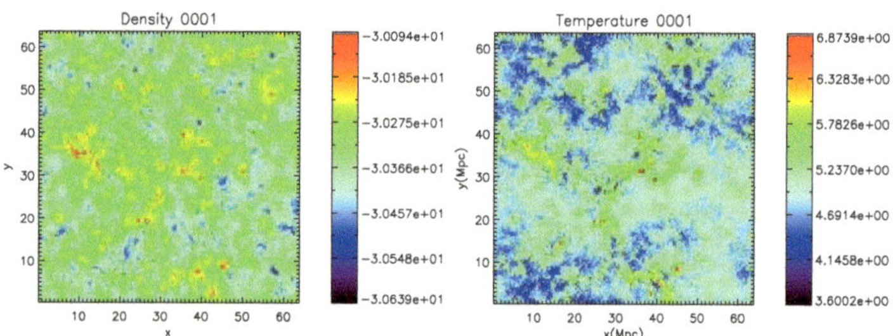

Figure 4: Log of gas density at time t=1.68e15 s

Figure 5: Log of gas temperature at time t=1.68e15 s

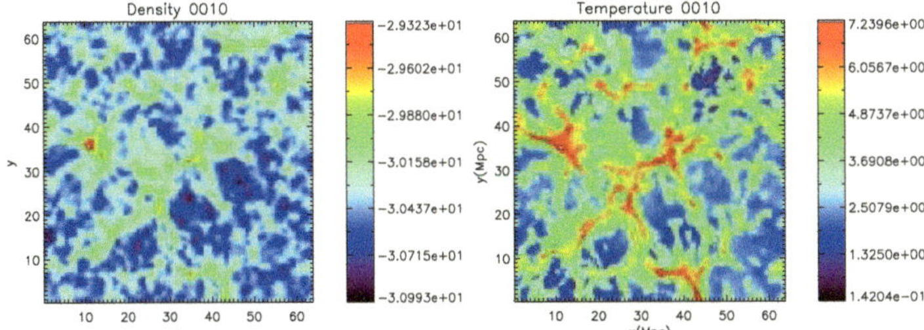

Figure 6: Log of gas density at time t=1.64e16 s

Figure 7: Log of gas temperature at time t=1.64e16 s

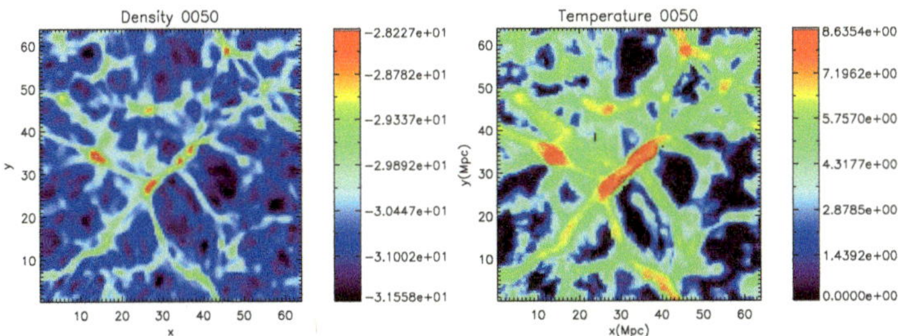

Figure 8: Log of gas density at time t=9.69e16 s

Figure 9: Log of gas temperature at time t=9.69e16 s

4.2 Evolution of the magnetic field

A correlation of the magnetic field and the density evolution through cosmic time can clearly be seen on the large scale. The regions containing high magnetic fields follow the shape and location of the high-density regions seen in the previous section (compare to figures 4, 6, 8 and 10). The plots also represent sections in the (x,y) plane through the middle of the simulation box. The comoving field strength increases in the region where the cluster is formed by up to 3 orders of magnitude as compared to the initial seed field.

The observed increase in magnetic field strength due to gas compression is expected due to the law of conservation of magnetic flux which is defined as

$$\phi_B = \int_A \vec{B} \cdot d\vec{A}$$

where $d\vec{A}$ is an infinitesimal area element. If we consider the gas to be contained in a structure of typical length scale L then $V \propto L^3$, where V is the volume of the structure. The density is inversely proportional to the volume, therefore $\rho \propto L^{-3}$ while the area is proportional to the square of L, $A \propto L^2$.

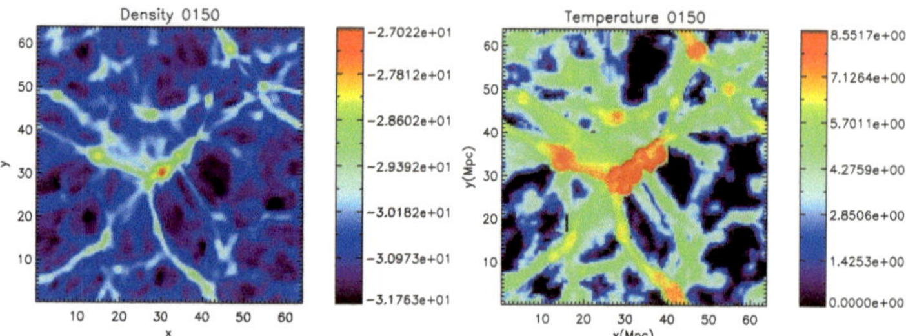

Figure 10: Log of gas density at time t=1.54e17 s

Figure 11: Log of gas temperature at time t=1.54e17 s

It follows that $A \propto \rho^{-2/3}$ and therefore the conservation of magnetic flux implies $B \propto \rho^{2/3}$, which explains the strong correlation between magnetic field strength evolution and structure formation.

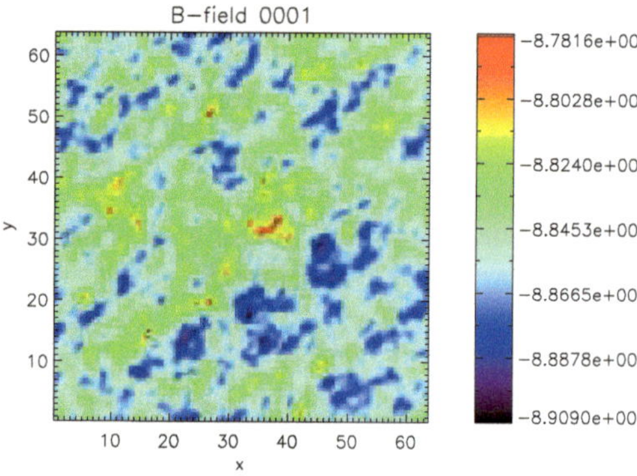

Figure 12: Log of magnetic field strength (in Gauss) at time t=1.68e15 s

To give a better picture of how the magnetic field evolves with time, the volume-averaged and mass-averaged comoving magnetic fields were calculated at each time step inside a box with side length 12.8 Mpc at the center of the simulation box (this is approximately the location where the galaxy cluster forms towards the end). Unfortunately the redshift could not be read from the parameters file due to a bug in the code, therefore the time evolution of the magnetic field in physical coordinates could not be determined.

It can be seen in figures 16 and 17 that the mass-averaged field increases from 0.1 to 35 nG while the volume-averaged field increases only from 1 to 3.5 nG. The difference between these two ranges reflects the fact that the

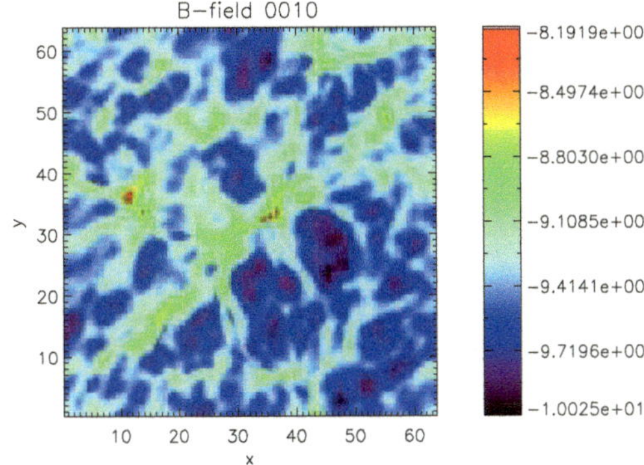

Figure 13: Log of magnetic field strength (in Gauss) at time t=1.64e16 s

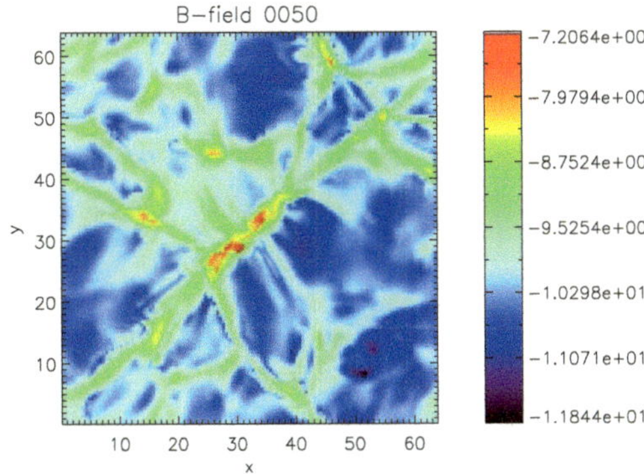

Figure 14: Log of magnetic field strength (in Gauss) at time t=9.69e16 s

density and the magnetic field strength are well correlated, which makes the mass-averaged field larger. Both the mass-averaged and the volume-averaged comoving fields increase steadily with time. An exception is the first time-step in the volume-averaged field evolution; in this time-step the initial uniform field was mixed randomly by the gas leading to a decrease in the volume-average. Moreover, in the mass-averaged field evolution we notice two peaks at about 1.2e17 s and 1.5e17 s. These peaks are very likely associated with mergers between the smaller subclusters and the central cluster.

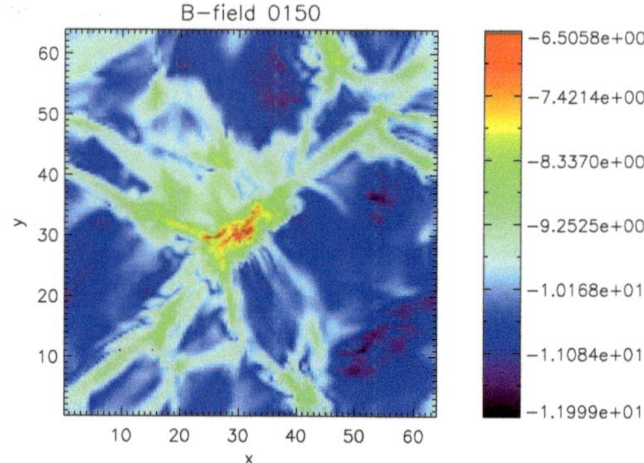

Figure 15: Log of magnetic field strength (in Gauss) at time t=1.54e17 s

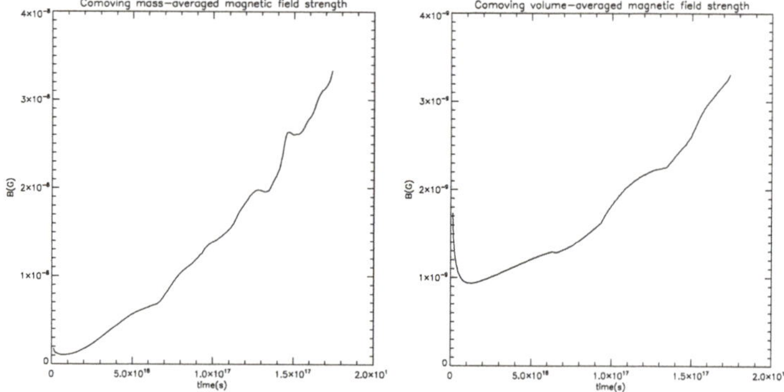

Figure 16: Time evolution of the mass averaged magnetic field strength

Figure 17: Time evolution of the volume averaged magnetic field strength

4.3 Properties of the final cluster

We further focus our analysis on the galaxy cluster contained in the most recent simulation file corresponding to a time of 1.74e17 s and a redshift of about 0.7. For this purpose, only a 12.8 Mpc box at the center of the simulation box was taken into consideration, since the cluster was identified to be contained within this box. Plots of the cluster density and magnetic field strength are presented below in (x,y), (x,z) and (y,z) sections through the cluster center. As can be seen from the density plots, the cluster is not yet fully relaxed, since it is clearly not spherically symmetric and infalling subclusters are still visible.

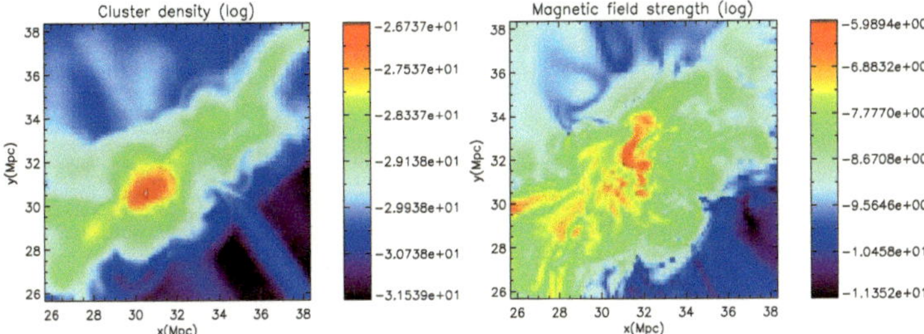

Figure 18: Cluster density, xy plane

Figure 19: Cluster magnetic field, xy plane

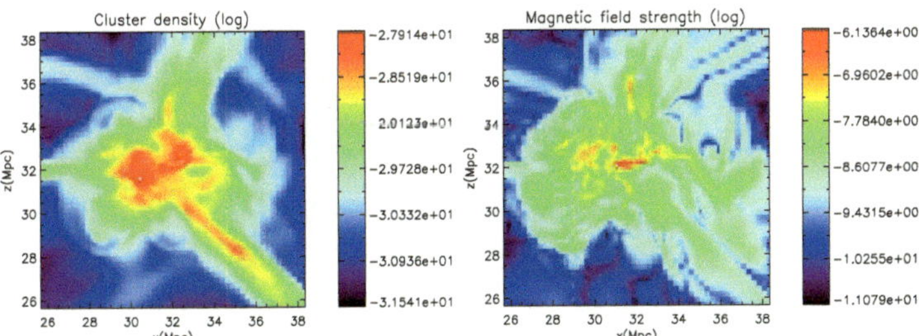

Figure 20: Cluster density, xz plane

Figure 21: Cluster magnetic field, xz plane

4.3.1 Correlation of the density and the magnetic field strength

As expected from observational data, the magnetic field is strong at the cluster center and decreases towards larger radii. However, the field shows considerably more substructure than the density maps and the maxima appear in patches distributed through the cluster center. A good visual example of this is a plot of the magnetic fields strength in the (x,y) plane through the center of the considered box (Figure 19), but both the (x,z) and (y,z) sections also show similar behaviors. In order to analyze the amplification mechanisms for the magnetic field we furthermore plotted the field strength divided by the 2/3th power of the density. If the fields were amplified by compression alone, a constant value of $B/\rho^{2/3}$ is expected. As this constant value was not obtained, we conclude that the magnetic field is also amplified by other processes such as shocks and shear stress. In fact, the largest amplification factors are usually

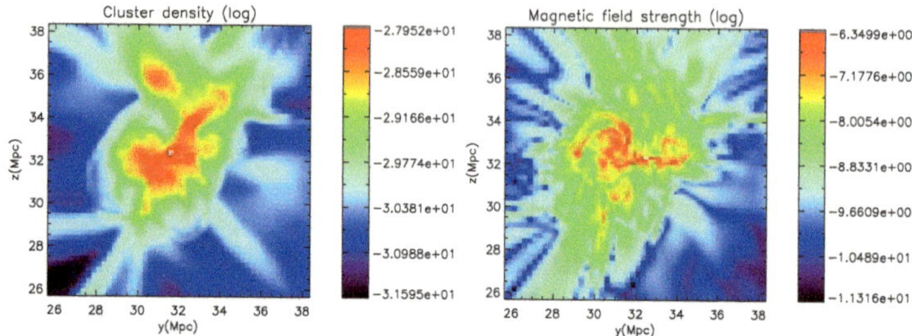

Figure 22: Cluster density, yz plane

Figure 23: Cluster magnetic field, yz plane

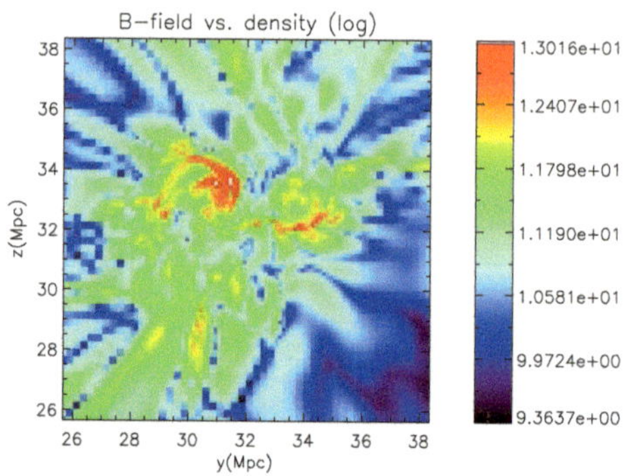

Figure 24: Magnetic field strength divided by the 2/3 power of the density, yz section

found towards the outskirts of the cluster where merger events which generate such shocks and shear flows have the largest influence. In Figure 24, the $B/\rho^{2/3}$ map is illustrated in the (y,z) plane. Comparing this to the density map in Figure 22 we notice that the largest amplifications correspond to the region between the in-falling subcluster (at the top-center of the density map) and the main cluster itself. This supports the argument that cluster mergers have a large influence on the evolution of the magnetic field. Moreover, the fact that the fields in the outer regions of the cluster are amplified higher than the fields in the cluster center confirms the observations of Dolag et. al. [10].

4.3.2 Power spectrum of the magnetic energy

Turbulence associated with gravitational structure formation (also known as Kolmogorov turbulence) shows a characteristic energy power spectrum of index -5/3. Ensslin and Vogt [6] found this to be in agreement with the observations of the energy spectrum of cluster magnetic fields. We plotted the energy power spectrum of the magnetic field in the simulated cluster and found the obtained slope to be very close to the expected -5/3, indicating that the magnetic field is indeed amplified by Kolmogorov-type turbulence. In the shorter wavelength range (for k > 10/Mpc, corresponding to length scales of 100 kpc), the spectrum deviates from the assumed power law due to the resolution which does not allow to accurately compute the effects of small-scale turbulent eddies. One advantage of the current simulation is the use of the piecewise parabolic Riemann solver to compute the evolution of the gas. This likely lead to obtaining the correct Kolmogorov spectrum, which was not the case for previous SPH simulations ([9], [10]).

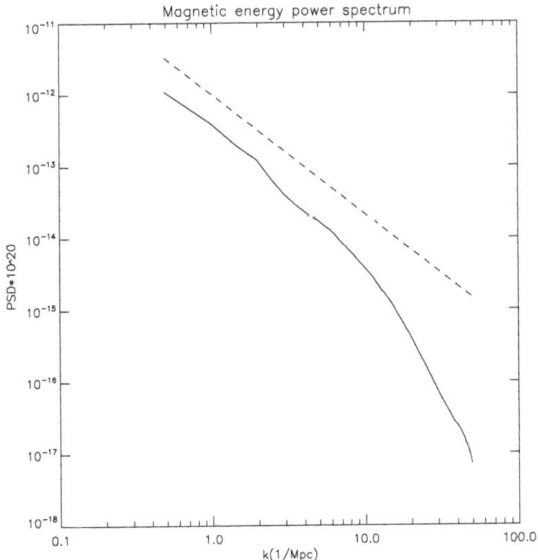

Figure 25: Magnetic energy power spectrum; the dashed line represents the -5/3 Kolmogorov spectrum for comparison

Moreover, we varied the size of the box centered around the cluster to side lengths of 7.68, 10.24, 12.8 and 15.36 Mpc respectively and determined the power spectrum in each case in order to assess box-size related artifacts. We found such artifacts to have very little influence on the slope of the power spectrum, the only quantity which changed owing to the box-size variation being the amplitude of each mode. This increased with decreasing side-length by a constant factor due to the exclusion of low-magnetic field regions from the box.

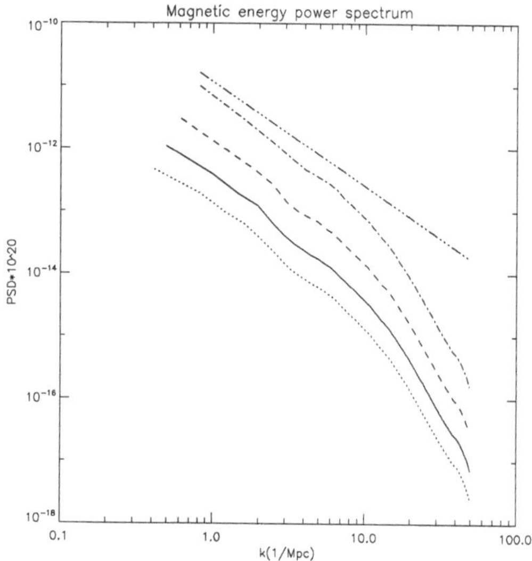

Figure 26: Magnetic energy power spectrum for various box sizes (dotted line: 15.36 Mpc, solid line: 12.80 Mpc, dashed line: 10.24 Mpc, dot-dashed line: 7.68 Mpc); the double dot-dashed line represents the -5/3 Kolmogorov spectrum for comparison

4.3.3 Rotation maps

Since Faraday rotation maps are currently one of the most important observational tool for determining cluster magnetic field strengths, we also calculated the rotation map for our simulated cluster according to the definition which can be found in equation (4). The integral was replaced by a sum of array elements multiplied by the physical distance between two such elements. An example of a rotation map projection along the x direction is presented in Figure 27 below.

The result shows, as expected, a zero background (compare with the density plot in Figure 18 for cluster location). In the cluster region, we observe the typical features expected of a rotation map, most importantly reversals from positive to negative values in neighboring patches. However, given our effective resolution, the distance between two neighboring data points was 62.5 kpc. Since the observed correlation length of rotation maps is of the order of 10 kpc, a better resolved simulation is needed in order to quantitatively discuss the artificial rotation maps obtained.

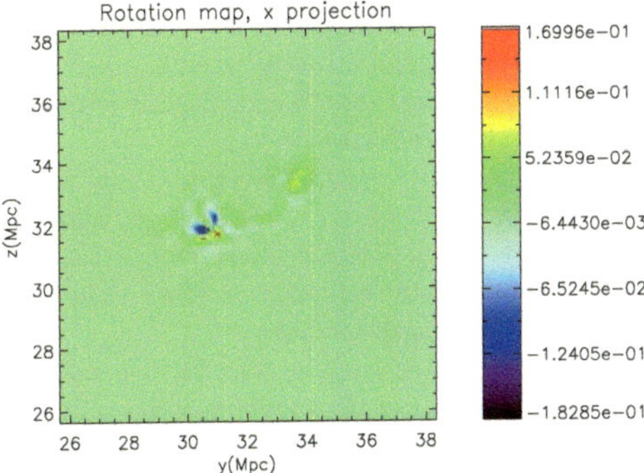

Figure 27: Rotation map, line of sight assumed in x direction

5 Summary and Outlook

Galaxy clusters are the largest gravitationally bound objects in the Universe. The intra-cluster medium has recently been discovered to be penetrated by magnetic fields with microgauss average strength and correlation lengths of the order of 10 kpc. We used the Flash 2.4 code to perform an adaptive mesh refinement simulation of galaxy cluster formation, with the purpose of testing how the magnetic field evolution is related to structure formation and whether the magnetic properties of the obtained cluster correlate well to recent observations.

The results of the simulation lead us to the conclusion that there is indeed an intimate connection between the evolution of large-scale magnetic fields and structure formation. The initial seed fields of 1 nG were amplified by up to 3 orders of magnitude during the formation of the cluster, and regions of high magnetic field strength can be seen to correspond to regions of high density. As expected from observations, the field strength decreases with increasing distance from the cluster center. Moreover, the mass-averaged magnetic field is an order of magnitude larger than the volume-averaged magnetic field throughout the simulation, also pointing towards the correlation between density and magnetic field strength. The power spectrum of the cluster magnetic field was found to be well described by -5/3 Kolmogorov-type turbulence associated with gravitational structure formation, another argument in favor of magnetic field amplification by galaxy cluster evolution. This characteristic of the power spectrum was also confirmed by processing of observational data.

Of the processes associated with structure formation, mergers seem to have an important influence on the magnetic field strength. This can be seen in the form of peaks in the mass-averaged time-evolution of the magnetic field as well as in maps displaying the ratio of the field strength to the 2/3 power of the density. The highest amplifications of the magnetic field relative to the gas density are seen to correspond to active merging regions towards the outskirts of the cluster. These results are in concordance with previous numerical simulations which used different types of code.

Realistic Faraday rotation measures were obtained from the simulation data, which was however not resolved well-enough to allow for a more quantitative analysis. To this end a better-resolved simulation is necessary, which would permit the more accurate computation of rotation measures as well as a more detailed correlation length analysis.

Further possible improvements of the code include solving the full magnetohydrodynamic equations rather than involving the magnetic field passively in the code and investigating other mechanisms which can enhance the cluster magnetic field, such as active galactic nuclei.

References

[1] T.E. Clarke, *Faraday Rotation Observations of Magnetic Fields in Galaxy Clusters*, Journal of the Korean Astronomical Society, 2004, **37**: 1-5

[2] L. Rudnick, *Observing Magnetic Fields on Large Scales* Journal of the Korean Astronomical Society, 2004, **37**: 1-5

[3] C.L. Carilli, G.B. Taylor *Cluster Magnetic Fields* Annu. Rev. Astron. Astrophys., 2002, **40**: 319-48

[4] F. Govoni, L. Feretti, *Magnetic Fields in Clusters of Galaxies* astro-ph/0410182, 2004

[5] C. Vogt, T.A. Enßlin, *A Bayesian view on Faraday rotation maps - Seeing the magnetic power spectra in galaxy clusters* A&A, 2005, 434, 67-76

[6] C. Vogt, T.A. Enßlin, *Measuring the cluster magnetic field power spectra from Faraday rotation maps of Abell 400, Abell 2634 and Hydra A* A&A, 2003, 412, 373-385

[7] R.M. Kulsrud, R. Cen, J.P. Ostriker, D. Ryu, *The Protogalactic Origin for Cosmic Magnetic Fields* ApJ, 1997, 480:481-491

[8] R. Banerjee, K. Jedamzik, *Are Cluster Magnetic Fields Primordial?* Phys. Rev., 2003, 91, 25

[9] K. Dolag, S. Schindler, F. Govoni, L. Feretti *Correlation of the magnetic field and the intra-cluster gas density in galaxy clusters* A&A, 2001, 378, 777-786

[10] K. Dolag, M. Bartelmann, H. Lesch *Evolution and structure of magnetic fields in simulated galaxy clusters* A&A, 2002, 387, 383-395

[11] C. S. Frenk, S. D. M. White, P. Bode, J. R. Bond, G. L. Bryan, R. Cen, H. M. P. Couchman, A. E. Evrard, N. Gnedin, A. Jenkins, A. M. Khokhlov, A. Klypin, J. F. Navarro, M. L. Norman, J. P. Ostriker, J. M. Owen, F. R. Pearce, U.-L. Pen, M. Steinmetz, P. A. Thomas, J. V. Villumsen, J. W. Wadsley, M. S. Warren, G. Xu, G. Yepes *The Santa Barbara Cluster Comparison Project: A Comparison Of Cosmological Hydrodynamics Solutions* ApJ, 1999, 525, 554-582

[12] *FLASH User's Guide: Version 2.4* ASC Flash Center, University of Chicago, 2004

[13] G. Tóth, *Computational Magnetohydrodynamics*, internal course notes, Eötvös University, Budapest, Hungary, 1998

[14] F.H. Shu, *The Physics of Astrophysics, Volume II: Gas Dynamics*, University Science Books, California, 1992